A New Vibrationary Model
of the Universe

**Quantum Alignment & Fracturing At the End of Time &
Quantum Lattice Structures Influence on Curves in Light
and Electron Orbits, how lattice vibration might explain gaps
in general relativity.**

by

Gianna Giavelli

first edition

To Kip, Stephen, Roger, and Kenneth

Chapter 1 – Lattices and Time

We think of the big bang as the reversal mechanism of the black hole star collapse on a grand scale, the emergence of singularity. What is missing is a picture of the long dead universe. And how might life and time re-emerge from that vast dead state. If you accept the expansionary model of the universe and increasing of entropy and congruent decrease in temperature then there must be an end state of unified no activity and low energy. Some argue for a re-contraction of the universe taking into account dark matter. But what if that never happens?

We have some theory for near initial state of the universe high energy such as the reduction of quanta into unbound stringlets of energy. But physics is wholly missing a theory for endstate other than simply stating it is a vastly expanded low energy equilibrium. And thus the question arises, how might anything occur from this "deep freeze" which is in effect another form of singularity where time stops. Rather than the unipointal singularity of a black whole, this is an all expanded singularity encompassing everything with zero state change. Energy is conserved but expanded to infinity such that it is nearly a no energy state at any one point. From this vast deep cold stop of everything what can emerge and how?

If you look to other cold systems nature might provide a clue. If you have ever seen pictures of arctic ice shearing off a massive glacier you realize that these shears occur because the nature of frozen ice is essentially crystalline and aligned. A near perfect face of ice falls off producing an explosive energy. Similarly, at end of time conditions, atoms may slowly lose energy and drift apart into quantum particles, quantum particles slowly distribute until all quanta have thoroughly expanded filling the void with a kind of "dust" that over aeons slowly rotate and align into perfect form. Quantum particles which engage the forces rather than ping back and forth come to rest in the middle. At the end of this mechanism you are left which something resembling a perfect crystalline matrix. Freeze ice quickly and you produce a cloudy mass. Freeze ice very slowly and you get a perfect glass like structure. Similarly, at the end of time, the universe resembles a vast window utterly transparent frozen at the slowest rate possible. In this lattice of quanta any rogue particle simply passes through and transmits without disturbing it. There are no collisions to re-propel the mass backwards, nothing to rekindle the spark of god.

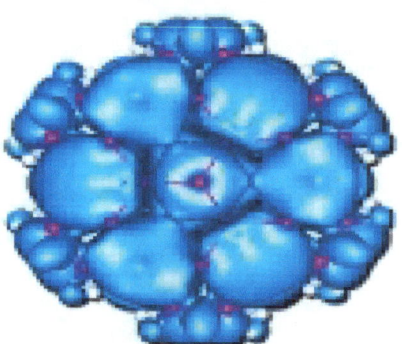

Quarks arranged in a very unusual shape – Quantum Lattice

After a vast amount of time, this steady state may be interrupted by something quite dramatic. The planes of quanta may shear due to minor differences in distribution and weight just as ice shears from a cliff. Slowly one plane, a billion quadrillion light years square, may begin to shear and shift away from the rest of the universe. Containing what was once a trillion galaxies but now appearing as void, this slow drifting shear may detach an even larger shear. This section of the seemingly dead universe suddenly springs to life in a quite odd way as more imbalance now arises gravitational effects. These sheared edges if the great expanded universe now undergo a transmutation like a decaying star but never are stars formed. Instead, quantum and stringlets begin to move closer and closer together and form a secondary complex structure which could never exist in any star's madness. In this structure they quanta and stringlets occupy overlapped space-time and like a crystal slowly cooled they produce a structure of a very different and perfect density unlike what we have ever seen in our universe and thus have no words to being to describe. A diamond lattice of quanta.

This perfect crystalline density of quanta has a very unique property above and beyond a singularity which is simply a stopped point in space time.

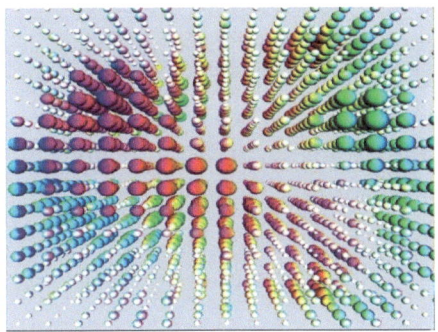

A diagram of ultra-cold atoms in a lattice formation

The nature of this is one of rotation. Movement. Movement at a vast rate. The reason this occurs is derivative of the earlier crystalline lattice perfect alignment. Rather than forming a banging ball of energy and gravity pulling things into a solar furnace, we get a cold collapse without the friction of that irregular state that we observe in matter agglomeration into a star or star collapse in our visible universe. We don't have a mathematics to describe this because again we are theorizing extra-universal phenomena. The energy-time of a quantum shear off the universal lattice at a size of $1/10^{th}$ of the expanded universe now being moved quickly into one small area of space-time, somehow cannot overcome the greater vastness and steady state and perfect formation of the end time universe. Thus the energy-time has a point where it bursts through, into a new space-time, into a new Fiat Lux movement of dramatic energy-time expansion into a void area. Boundless areas of pure voids overlap with boundless areas of active universes and other boundless areas of end-time universes. It is the explosion of TIME from the void of seemingly nothing, from a state of perfection in structure some will call God.

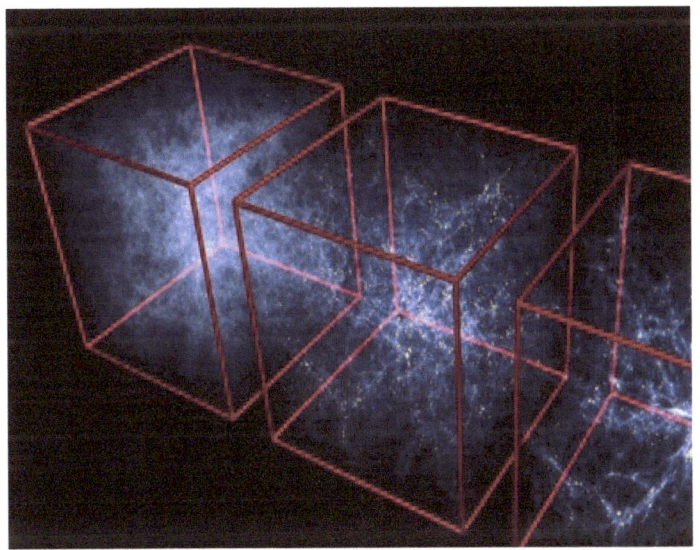

Diagram – Universe moving to pure distribution at end-time

The physicist and cosmologist does not say God, instead she says instead quantum chromodynamic lattice structure. It sounds more edifying at night.

Now we are left with the dubious question of deriving a mathematics for the unobserved universe and unobserved states. Is such a task even possible? How can such a proof of this system be created. And there is one answer. Presently we believe that a big bang is essentially a chaotic event and that matter clumps randomly into clusters and super clusters in space. However, if there is a lattice structure at the other end of time then we should see withing our own universe hints of that structure.

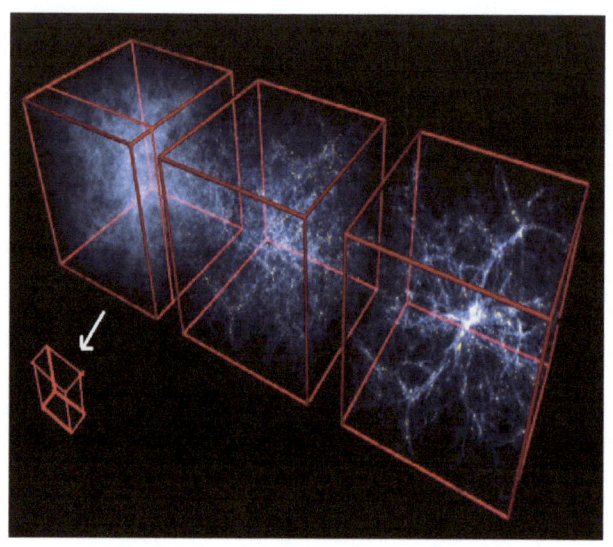

Diagram – Cleaving of Lattice Space-

Can very different universes emerge which run at different
time speeds? Are we in a hypersonic universe or a painfully
slow one all the others blurring out in our seconds?

Can we picture it as a concert with each universe's expansion
and deflation all occurring side by side making such wonderful
noise?

Chapter 2: The Universality & the Pasta Extruder

So what is a universality? The opposite of a singularity in all regards. A vastness where space-time has been propelled into rotation. Does a diamond lattice structure of quantum/stringlet exist in our low energy disordered universe? Can it be lab created? Do they shear? More questions in an endless quest for understanding of how we begin and how it ends, or more precisely, how it never ends.

When you make pasta, you use different shaped dies that look nothing like the end product noodle, and at high pressure force the dough through it. Then depending on the speed of rotation of the pasta cutter blade, thousands of different end shapes are possible. If you understood the science of pasta making, you could look at the noodle and know exactly the shape of the die, and the speed of the rotation of the cutter. Similarly, as our knowledge of the early universe increases we may seek our clues, except it is the speed of time that is our blade.

Now, this begs the question: What is the PRESSURE of our universe. If one universe can burst through to another this must be a function of pressure.

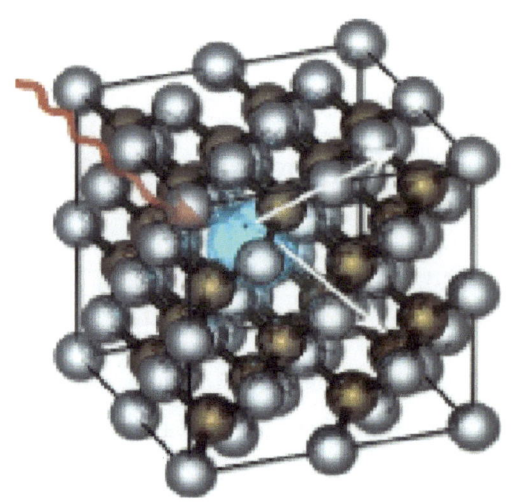

A diagram of aluminum antimonide, a lattice structured molecule, is a potent gamma radiation detector. Anything entering the lattice (red) disturbs the electrons (blue)

If we have an adjacent universe which has achieved unfathomable high pressure, what would it be if it rended our fabric of space-time and burst in? Because we are a non-empty universe, then this would not be a universe creation event but simply a stream like air running out of a tire leak. A high pressure energy-particle stream that has rotation (because the collapsed universe is rotating?). Where do we see evidence of these "White Streams" ? There is only one. At the very dawn of our universe strange objects did appear and because of the nature of light and distance we can still see them today. They are the most far objects we can see. We call them Pulsars. To science they are a mystery. But what if in the early fabric of our universe we did NOT have sufficient pressure to prevent breaching into our space? The result would be multiple leaks into our universe from tiny streams. The skin on an egg shell stretched too thin. As our universe stabilized in pressure and expansion, becoming less than void, leaks would disappear and we do not see such things in our modern current universe.

Chapter 3 – Staring for a Long Time At An Egg

What happens when you consider quantum theory is that suddenly the universe becomes strict. Perhaps too strict. We suddenly had a very strict "Particle" theory of the universe. Is this correct? I think not. We got there by smashing things into smaller and smaller pieces. Atoms begot quarks begot leptons begot mesons and pions.

Take a log and burn it. You are left with a pile of soot. Ahh the log is comprised of SOOT! You scream and marvel. Yet look at a slice of the log under the microscope. You see vessicles transporting fluid up and down the tree. You begin to understand that you are looking at a transversal slice of a pump – in this case for sap. Now look at the gray ash. Its all small particles. So, the quantum theorist boldly exclaims AHA the log is truly a GRAY thing made up of the finest particles. I have a distinct sense that the quantum theory which materialized from atom smashing experiments there can be.

Roger Penrose, the theorist who gave us the black hole recently criticized Quantum theory the following way -

> *Quantum mechanics is an incredible theory that explains all sorts of things that couldn't be explained before, starting with the stability of atoms. But when you accept the weirdness of quantum mechanics [in the macro world], you have to give up the idea of space-time as we know it from Einstein. The greatest weirdness here is that it doesn't make sense. If you follow the rules, you come up with something that just isn't right.*

> *In quantum mechanics an object can exist in many states at once, which sounds crazy. The quantum description of the world seems completely contrary to the world as we experience it.*
> *It doesn't make any sense, and there is a simple reason. You see, the mathematics of quantum mechanics has two parts to it. One is the evolution of a quantum system, which is described extremely precisely and accurately by the Schrödinger equation. That equation tells you this: If you know what the state of the system is now, you can calculate what it will be doing 10 minutes from now. However, there is the second*

*part of quantum mechanics—the thing that
happens when you want to make a measurement.
Instead of getting a single answer, you use the
equation to work out the probabilities of certain
outcomes. The results don't say, "This is what
the world is doing." Instead, they just describe
the probability of its doing any one thing. The
equation should describe the world in a
completely deterministic way, but it doesn't.*

To me, this sounds like quantum mechanics is a incorrect
system trying to predict a flowing system in motion, a system
that is fundamentally fluid. The real dynamics of atoms seem
to me to be more of vibration and spin, spinlets of energy
which buzz. Place a pack of peas on a vibrating plate. Now
close your eyes and try to feel them. They will mush in and out
without form. I wonder if the true structure of atoms isn't more
like this. Except instead of hard peas they are squishy koosh
balls with increadible levels of squishyness.

If we could take a simple atom – Hydrogen – and make it as
big as our head, what would we see? Would we see an electron
going around a protron ? Is Bohr right? He is certainly
accepted as fact in our world.

Bohr Model

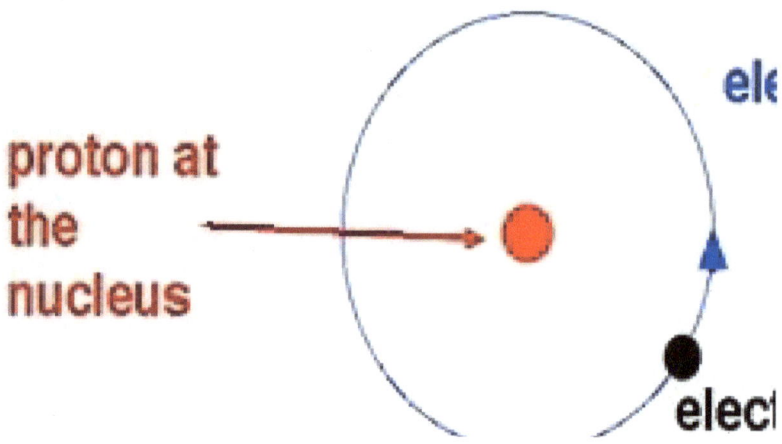

proton at the nucleus

Or, did we, with so much success with Kepler and Copernicus and Galileo determine that As Above So Below the inner world must model the same physics as the grand scale systems like our solar system.

In the probability model, the hydrogen atom is more a squishy ball.

What if, there is no electron? What if it's simply a collection of dust. In 1926, Erwin Schrodinger used this idea to develop a mathematical model of the atom that described the electrons as three-dimensional waveform rather than point particles. A consequence of using waveforms to describe particles is that it is mathematically impossible to obtain precise values for both the position and momentum of a particle at the same time

The Bohr planetary model was rejected for one of electron orbit. So why did we go back to particle models for quarks? Our mechanics of sub-atomic VIBRATION is truly the new physics that is much deeper to think and fathom and harder to conceptualize, and it is very different from quantum theory. Vibe-atomic theory might go something like this – the Universe operates at a resonance frequency. A resonance in the fabric of space-time. As we get closer to the fabric of the universe we see the resonance effects of different vibratory systems. If we could hear the universe, it might to use sound like a choir of angels. Revolving spinning and twisting bundles of energy not particles. The same potential energy wells that describe electron position and orbitals could be worked out as spin and vibration if we put our heads to it.

And now we come to our prime topic. Take a atom of Helium, a supposed stable atom, and put it in a box. Now stare at it for a trillion trillion years. What happens? What happens to the neutron, the protron, the electron? We know the motion is not perfect and even pendular motion systems lose speed over time. Every clock winds down. So why not with these energy systems as well. What is the nature of the un-wound atom? A pile of strings from string theory? Or does vibration simply slow until energy dissipates as radiation? Like a magicians trick, at the end, there's nothing there at all. It is the vibration itself, not as a gravity function but constrained by atomic gravity, that provides the illusion of particles. Is this saying anything new? This has been pointed to but maybe not as directly. If we can take the "stable" helium atom and put it through time travel, and take a picture of it every billion years, then replay the minute long movie, we might see an astounding picture of micro explosions in matter. Maybe the unwinding is dramatic? Maybe it's imperceptible. Both are good questions. If truly quantum physics is right and the particlists prevail, at the bottom of our box we should find a bunch of quantum particles just sitting there exhausted. What is the experiment to test this? Our one advantage is that there are so many atoms we might catch a glimpse of one in this end state. But if you consider the energized birth of atoms in our stars, this energy boost from the initial might last ten billion years, time much longer than we can ever see one observable.

Chapter 4 – Electron Orbital Decay and Synchrotron Radiation as a Clue to Unified Theory

Electrons orbiting a nucleus should lose energy to radiation and eventually spiral into the nucleus. This is not observed. Atoms are stable on timescales much longer than predicted by the classical Larmor formula.

Electron capture is a process in which a proton-rich nuclide absorbs an inner atomic electron, thereby changing a nuclear proton to a neutron and simultaneously causing the emission of an electron neutrino. Various photon emissions follow, as the energy of the atom falls to the ground state of the new nuclide. Electron capture is a common decay mode for isotopes with an over-abundance of protons in the nucleus. What is interesting about the phenomenon of electron capture is that it depends not on the electrons in the electron cloud of the atom, but rather on the nucleus. Thus, one can not ignore the fact that the behavior of electron capture is dependent solely on the nucleus, not the electrons.

For example, if the nucleus is, for example, Carbon-9, 100% of this isotope will decay via electron capture to 9-Boron. Yet Carbon-14, which has the same electric charge and same number of electrons in an identically configured electron cloud, never decays via electron capture. Quantum physics, especially when the answer is focusing on the electrons of the atom, has trouble explaining the behavior of Electron Capture with a sufficient credibility.

What if there is no motion of an electron. What it it is instead, and epi-phenomenon in space-time. As the nucleus vibrated and moved it simply beamed an effect of negative charge around its center. Much more complex larger nuclei would beam this effect into multiple points that wobble, and are resolved as our electron orbits. The electron orbit decay to a lower state releasing a photon would instead be a release of energy from the nucleus. Think of them like circular beams from a spotlight at a giant show. The spotlight wobbles and the beam appears to move around.

What about electron decay converting a protron to a neutron and releasing a electron neutrino? What if this instead was more akin to one of the six bulbs in the spotlight burning out. The protron itself changes and releases the "electron neutrino"

If we envision the electron as a bundle of energy that spins (thus generating a negative charge) if you remove the spin then perhaps you would have a neutrino, a bundle of near mass-less energy without charge.

This is something that deserves research and prodding and poking to arrive at a non-quantum model. It is unifying in that we observe similar energy emission in the very vast relativity scale, as exhibited by black holes and nebulae. By thinking past quantum and particles and thinking more in terms of excitation to spin bundles of energy with some spins being stable, others rotating, and other violent (wobbling twisting) we might better see this universe not as units of particles but as twists of energy-lets.

Crab Nebula. The bluish glow from the central region of the nebula is due to synchrotron radiation.

The same radiation is observed at the large scale in black holes as they twist through space. "by ejection of jets produced by gravitationally accelerating ions through the super contorted 'tubular' polar areas of magnetic fields. Such jets, the nearest being in Messier 87, have been confirmed by the Hubble telescope as apparently superluminal, traveling at 6×c (six times the speed of light) from our planetary frame. "

It seems that such synchrotron radiation should not occur if you believe in a perfect void of space. It is also untirely unclear in physics the mechanism by which a magnetic field induces such effect on accelerated particles traveling in orbits. There is a gap in physics and positing tiny quanta exchange as energy exchange seems to be missing the ability to truly describe many phenomenona in our universe.

I believe these two problems are related – the complexity of electron orbitals and why they don't decay quickly, and why jets in space exhibit synchrotron radiation.

Chapter 5 – The Speed of Light Revisited

Humans can see light from 300 to 1000 nanometers in wavelength. The speed of light is a constant. Why? Why is it a constant. Physicists and Astronomers reply "because that's the fastest anything can go". This smacks very much up against our human experience. Make a race car and soon someone will build a faster one. With more energy speed increases. The 1000 horse power Bugatti beats the Porsche.

For centuries physicists struggle with the dual nature of light without explanation. At time it appears as a wave and other times it appears as a particle.

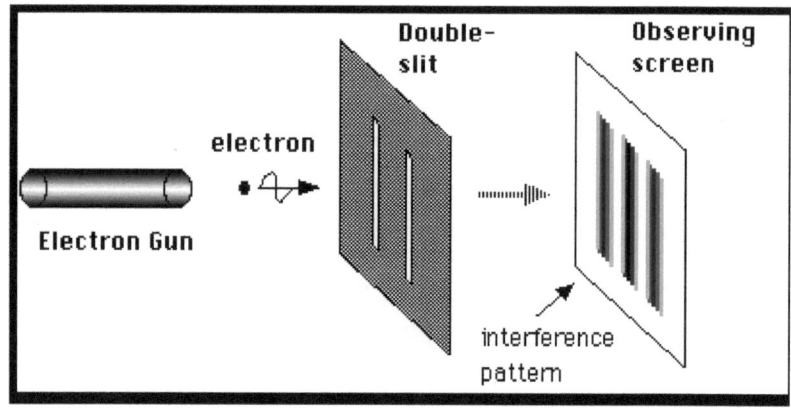

Young's experiment or the dual slit experiment showed that coherent light or electrons produced a pattern of smooth graduation in intensity up and down with an interference pattern such that each source was additive and subtractive. This can only happen with a wave. But the odd thing is, this was happening not just with light, but with particles, electrons, and even very heavy molecules consisting of 810 atoms.

When I was first taught this in junior high, the teacher explained to me that this only happened with light. So I thought aha it's a weirdness of light due to it's tiny size and tiny mass. But upon hearing that no, it seems to be for ALL particles in our atomic dictionary, that thrust me in the direction that there is something so fundamental in nature that physics has ignored. The problem is that physicals simply says "Oh its the dual nature of particles that defeats the classical model" and stuffs itself up quiet. That simply isn't an explanation at all.

Now what if we think of that same particle, but traveling through something which is effecting it. Faraday proposed and discovered that polarized light the direction of the polarization could be rotated by applying a magnetic field. *luminiferous aether* proposed by Huygens in 1678 was cast into strong doubt in the late nineteenth century by the Michelson-Morley experiment. The device he designed, later known as a Michelson interferometer sent yellow light from a sodium flame (for alignment), or white light (for the actual observations), through a half-silvered mirror that was used to split it into two beams traveling at right angles to one another. After leaving the splitter, the beams traveled out to the ends of long arms where they were reflected back into the middle by small mirrors. They then recombined on the far side of the splitter in an eyepiece, producing a pattern of constructive and destructive interferance whose transverse displacement would depend on the relative time it takes light to transit the longitudinal *vs.* the transverse arms. If the Earth is traveling through an aether medium, a beam reflecting back and forth parallel to the flow of aether would take longer than a beam reflecting perpendicular to the aether because the time gained from traveling downwind is less than that lost traveling upwind. Michelson expected that the Earth's motion would produce a fringe shift equal to .04 fringes—that is, of the separation between areas of the same intensity. He did not observe the expected shift; the greatest average deviation that he measured (in the northwest direction) was only 0.018 fringes(from wikipedia).

Later Einstein was awarded the 1921 Nobel Prize for his explanation of the photoelectric effect. Experimentalists had found that '... when a metallic surface is exposed to electromagnetic radiation above a certain threshold frequency (typically visible light), the light is absorbed and electrons are emitted. In 1902, Philipp Eduard Anton von Lenard observed that the energy of individual emitted electrons increased with the frequency, or color, of the light. This was at odds with

James Clerk Maxwell's wave theory of light, which predicted that the electron energy would be proportional to the intensity of the radiation. ...(Wikipedia)'

Einstein's theory, published in 1905, accounted for the emission of electrons from a surface, when it is illuminated by light, by describing light a a 'photon' or quantised packet of energy. The energy of the photon was described as being directly proportional to the frequency of the light. With the constant of proportionality, within the energy 'E' / frequency 'f' relationship, being Planck's constant 'h'. Thus, a photon of light has a 'quantised' energy given by the expression: -

$$E = hf$$

Now once again we ask WHY this is so.

So what we make of light, is that it is a kind of bundlet of energy (can we say particle?) that is moving along a sinusoidal trajectory through space-time and this trajectory can be influenced by magnetism. James Clerk Maxwell developed the mathemetics to unify description of el;ectricity, magnetism, and optics and later Einstein reified them into a basis for special relativity. Our experience of these things is unified in that we see they are related and effect on each other in mathematical harmony. But that doesn't give us the WHY of it.

And now we poke a stab at WHY the speed of light is constant.

Let's start with the creation of light. We spew off this bundlet of energy with unique characteristics we call a "photon". To us, a photon is electromagnetic energy within our visual range. Now let's say that initially this bundlet behaves the laws of classical motion and travels in a straight line. But something influences it. And the influence is that it starts to wobble and oscillate up and down until it's motion is effected. This wobble extends into a sinusoidal pattern the bundlet rising up and down. This bundlet has a frequency. For visible light it is **430– 790** terra-hertz or trillion times per second. Now we must return to the relationship of light and a stream. Light is not created by a particle splitting off a mass like a protron being ejected by the sun. No. Light is ejected as a stream continuous energy burst out. If you could look at the first meter of light, you might see that it is a straight line of perfect energy. But as it continues to move, something in our universe is limiting it. And this limit results in the distortion of light such that it oscillates, the greater the SPEED of the bundlet the more intense the oscillation. This makes the effective distance to be covered GREATER. If you have a sine wave of one meter in length and a particle were to traverse it, you would go about 1.5 meters in total distance to cross the meter of distance. Less efficient that straight. But if you had a thousand sine waves to traverse each a meter high, then the total distance to traverse is now 1,000 meters. So if the effect of the universe is to increase oscillation on the emitted bundle/stream of energy relative to its speed, the resulting effect is a constant speed of light. The truth is that the speed of light is NOT constant, only our measurement of its distance is inexact as we forget to measure it's amplitude from trough to hill as it moves up and down in wavelength.

Now, if we think again to this jet of energy called light which is continuous over time we might have a concept of a wavelet of light if light is only emitted for one millionth of a second. This tiny spaghetti noodle of light would move forward oscillating up and down exactly in proportion to its initial speed/energy. We as of yet have not done much experimentation into these snips of light on earth, as we do not have very good means of producing them but it would be a novel and unique science.

Now one question is does the jet of light as it travels into this wave patter separate into bundlets of energy such that larger masses of it exist in the sine =1 and sine =0 points on the line (or top and bottom of the curve). If the energy starts to shift and group, then we do get the expression of a machine gun of particles hitting something rather than constant energy. This rapid flashing of impact is exactly what our eyes were designed to perceive. When we get impacted by blasts more slowly of these machine gun bullets of energy, we see red. Faster and we see purple.

Now we can see why E=hf and why this must be so. Each bundlet of energy is actually the same, simply the higher velocity of the high frequency light has more bundlets of energy captured in more waves which hit at a faster rate.

Now we have a mechanism structurally to understand the dual slit experiment. Energy may be concentrated in the hill and trough of light, but some still remains throughout the squiggle, and this is what produces the perfect wave diffraction pattern. So we are looking at a continuous jet of energy, deformed by something into a sinusoidal pattern, with clumping at the extreme values of sine. This nature, increases distance with energy such that greater energy cannot produce greater speed in the macro observational but does actually achieve greater speed when all motion is accounted for.

Now somehow I am beginning to feel a bit more comfortable with the concept of a constant speed of light. I understand the mechanism a bit better of how the universe constrains it. But I am still at a loss for why, what happens to bend our perfect foot of spaghetti straight into a wavy noodle? What is this effect? This seems to be the big question they all overlooked.

There seem to be two possibilities. One is that there are built in micro-variances of magnetism in perfect symmetry across our universe. And these tiny areas pull and push on our spaghetti string of light until it deforms.

Another theory is the vibrationary model. Which is to say there is some kind of fundamental vibration to space-time. Perhaps the speed of the universe is related to the speed of this vibration and other universes might be created with different rates which allow for vastly different speeds. Now this effect would be similar to the magnetic variance except that there is no variance not banding of the field expression. The field expression itself moves during time.

This would then cause the jet of light to want to slowly move in harmony with it and oscillate as well. This effect would increase as the energy/mass of the bundlet was increase producing even more oscillation. And lower energy would produce lower oscillation. If this were reversed, it may have very well been then case that E = constant regardless of frequency not $E = hf$.

Whatever the effect is, it is pure, omnidirectional and uniform. This is why the Michelson-Morley experiment will never detect it. There is no "wind" and the speed of our planet moving through it does not change it or increase it. This may be an effect of our space-time substrate itself.

In the vibrationary model of the universe, this oscillation effects everything but in different observable scales. Something as large as our earth may have almost no resonance whatsoever. In fact, we only observe its effect on light because light is a continuous jet of energy. Because it is continuous we observe these tiny shifts and sine movements in it. And the squiggle of the sine is not at all uniform with respect to direction. Some are up and down, some left and right, some at 45 degrees. Perhaps polarized light will tell us more.

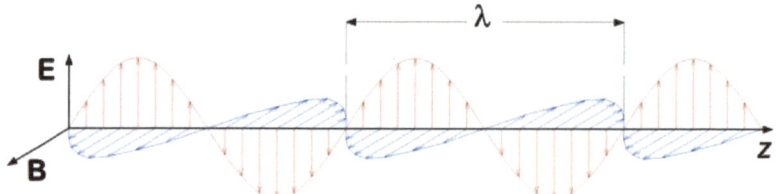

The magnetic field extends along B, orthoganol to the electric field which is traveling along E. In fact one might summise that the magnetic field is a side effect of E.

This diagram shows us the clumping of energy along the wave pattern:

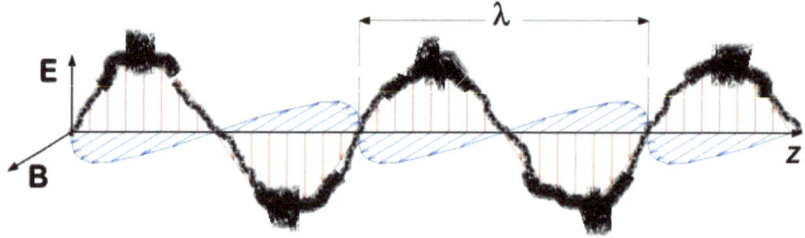

And you can just imagine that as it crashes into a wall it produces a kinds of drumbeat thump thump thump behind it's pattern.

If we could create a thin material with a billion billion holes, all exactly at a 45 degree angle, then only light which was perfectly hitting such that it was at the down slope of the cycle could get through. But all the light on the other side, a greatly reduced number, would all be traveling in perfect unison. Thus we have polarized light. And if we could put our microphone to it before and after, we might hear static, but the pure polarized light would sound like pitter patter perfect beats with perfect rhythm.

Chapter 6 – Vibratory Models on All Particles

Now if we know that the Michelson-Morley experiement held true for all particles and even small atoms, then perhaps the vibrationary effect which we so described with light is working for all particles as well, but in more limited ways.

As particles increase in mass, this effect may diminish such that oscillation and curve is either more limited or distorted. This may be precisely what is giving rise to electron orbitals.

For now, say that electrons were like bundles of energy sheared off the nucleus like a giant spotlight. Two protrons produce two spotlights. Four four, and on and on. This bundlet of energy might begin the same sinusoidal attempt as light, but now with much smaller space and mass, and gravitational pulls from the atomic center itself, bend inward on itself. So perhaps the electron is not so much a particle trying to follow an orbit (which must decay!) but rather, a side effect energy burst of a protron, now hopelessly bent by the effect of space-time. For Helium you get a perfect figure 8 pattern. And not having enough energy to escape this effect so it cannot burst forth as light. If you could somehow turn UP the volume on the spotlight emitting from the protron perhaps light would burst free? One wonders if the movement of the electron orbital is simply the wobble of the protron. That would be a good experiment to test if it were possible to detect protronic position vs. electron and induce spin on the protron. Somehow this notion that the electron is simply a shadow effect of the protron is seductive but not provable. But whether it is an effect/spotlight or a true bundle of freed energy doesn't matter, the effect governing its motion is the same, that is the same sinusoidal induced curve that light experiences. Because light is traveling with such greater velocity it linearizes and becomes a sinusoidal wave. Compressed, it is an orbit, zinging around endlessly. It is a shape destination caused by the vibrationary model effect upon the movement of all energy in the universe.

We do have one more particle we can look to and that is the nature of the escaped electron. We might think of an electron as being a bit like a bundle of energy like light. In atomic decay, sometimes an electron is emitted as a Beta particle/wave.

In water, beta radiation typically exceeds the speed of light in that material (which is 75% that of light in vacuum) and thus generates blue Cherenkov radiation when it passes through water
So light is not a different particle – the photon – it is an arrangement of phase of a bundlet of energy. And when it is within our receptors range, we see it as visible light.

We can prove this with an experiment. The **Smith–Purcell effect** was studied by Steve Smith, a graduate student under the guidance of Edward Purcell. In their experiment, they sent an energetic beam of electrons very closely parallel to the surface of a ruled optical diffraction grating, and thereby generated visible light. Smith showed there was negligible effect on the trajectory of the inducing electrons. Essentially, this is a form of Cherenkov radiation where the phase velocity of the light has been altered by the periodic grating.

Electrons turning into photons? How does this make sense? The diffraction grating was essentially scooping off clumps of energy in a wave pattern. Reconstructing a wave of light clumps of energy.

So why do we get Cherenkov radiation from faster than light radiation? Simple, there is enough energy there to allow some to be cleaved off into a light which we call Cherenkov radiation and while simply producing radiation isn't so spectacular what is so impressive about Cherenkov is that it is in our visible spectrum, a holy blue. Perhaps the only time that invisible radiation transmutes to visible light.

In this way, we unify the very vast – the travels of light and the crab nebula -with the impossibly small – the travels of the electron – with one theory. And one would ask well what experiment can we devise to prove this effect and all I can say is we already see it in the nature of light and all electromagnetic spectrum.

To make it mathematically sound, the effect of this vibration should be linear or exponentially symmetric as one scales energy or mass or speed.

We have some hints against space-time super symmetry with the Casimir effect. The Casimir effect is a small attractive force that acts between two close parallel *uncharged* conducting plates.

The attractive Casimir force between two plates of area A separated by a distance L can be calculated to be:

$$F = \frac{\pi\, h\, c}{480\, L^4} A$$

where h is Planck's constant and c is the speed of light.

If electromagnetism was supersymmetric there would be fermionic photinos whose contribution would exactly cancel that of the photons and there would be no Casimir effect. The fact that the Casimir effect exists shows that if supersymmetry exists in nature it must be a broken symmetry, or, a lattice.

Two plates a meter square at a distance of one micron, produce a Casimir force of approximately .1 grams. A real effect on our universe. (Philip Gibbs, 1997)

Is the Casimir effect a hint at the SQL structure? It begs the question what is the scale of this lattice and could you produce an experiment which shows a positive effect at one micron of space, but a negative effect at the next micron adjacent. Or does the SQL structure by its nature only produce positive effects? What if we could scientifically measure Casimir effect when the distance between plates were not micron but nanometers to approach wavelengths of light? Wouldn't it be remarkable if we found this differentiation in effect in a span equivalent to x-ray wavelengths?

Chapter 7 – Cherenkov Radiation as Mediator of Superluminous Speeds in Our Universe

In the beginning of this book we discussed quantum lattice frameworks. But the deep structure of space-time, normally invisible to us, is not quantum based. It is something much smaller. If we picture the fabric of space time in the end time unraveled universe, and conceive of the energy that once bound atoms together having dissipated, the energy that bound quarks together dissipated, the energy that bound quarklet-energy-strings together dissipated, then what would you have. This is the great unraveling of the universe at the end of entropy and enthalpy.

Let's refer to the structure of ice as a reference for the very cold. The normally polar molecule of water arranges perfectly in slowly chilled ice forming clear ice. If you were to slice it and look at a cross section you would see the Oxygen-Oxygen tip of molecule A, then the Hydrogen tip of molecule B, then the Oxygen-Oxygen tip of molecule C, then the Hydrogen tip of molecule D ad infinitum.

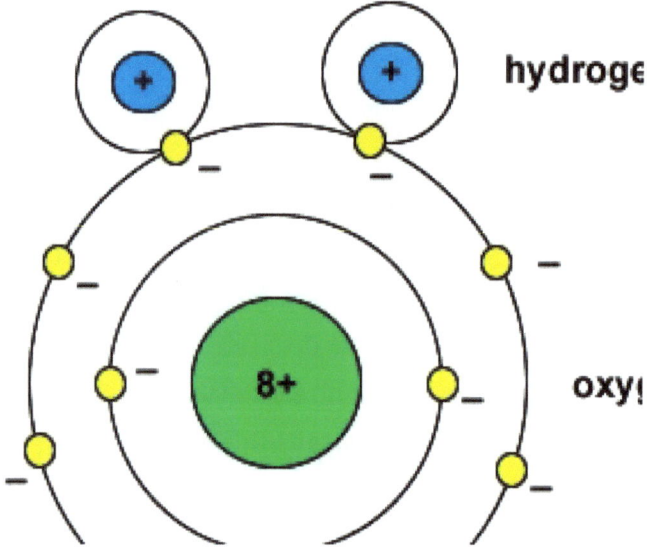

Skimming across this surface you would encounter

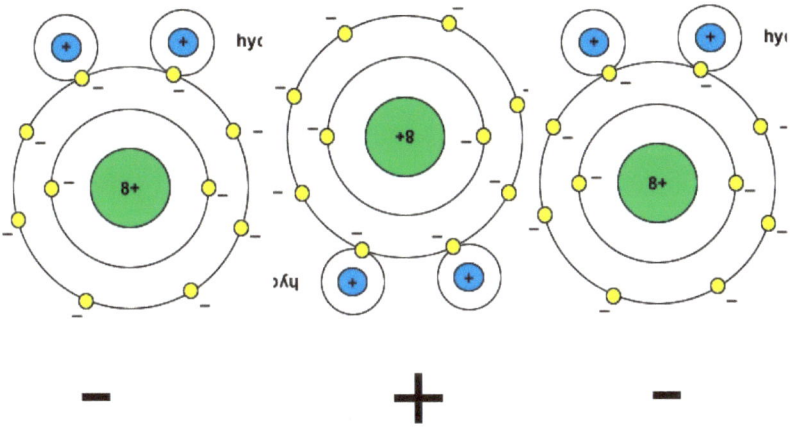

No this is water but if you adopt a lattice theory for the universe then this same state should exist for even a rough lattice of de-energized quark-stringlets-energies. If this is the structure of the universe then it might explain why we develop waves.

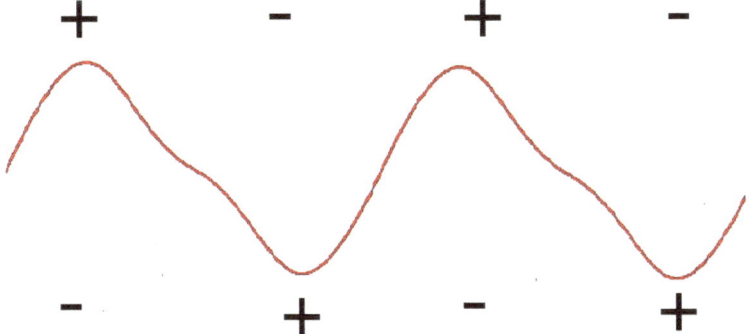

Each polar tip pushing and pulling on waves. For lower energy infrared it would skip say every hundred polarization points but still finally flip sinus position up and down. As more energy is forced through, ultraviolet then micro and xrays would be ejected as the wave frequency became higher and wave amplitude became shorter.

At a certain point, this system reaches maximum but there is still more energy in the particle. At this point there is some kind of shear effect akin to friction where the forward motion of the particle (not the wavelength) exceeds the speed of light (the natural balanced maximum) and energy is ripped away from the bundlet of energy as it pushes forward. This energy sheared off pulses in the up and down wavlet at a down-harmonic of the original and slowly removes energy from the system until it is no longer super-luminous. And this energy escaping is Cherenkov Radiation.

In all the places we witness such radiation we are dealing with ultimate high energy physics – fissioning nuclear cores, nebulae, and cosmic jets.

Cherenkov Radiation is the brake on the universe. It does not make so much sense to explain this braking action as particle exchange like the Quantumists would have us.

No, the Michelson-Morley experiment failed BECAUSE the speed of the earth is not super-luminous. Let's envision a better experiment, a Giavelli-Morley so to speak. How would we do it?

We would take our test apparatus and accelerate it to .9 speed of light. Then we would launch our photons at a speed which in one plane would EXCEED the speed of light such as 0.5 the speed of light. What would we expect?

The line traveling at 1.45 the speed of light would very briefly emit Cherenkov radiation as it is braked to the speed of light. The two beams would arrive out of phase due to the different speed at the very beginning of the pulse before it slowed.

Our universe seems utterly transparent to us because it is rare that we challenge its transmissive limits. All particles pass freely through it until luminous speeds are reached. Is this the photon simply increasing its gravitonic distortion as speed increases? Is space time warping as speed increases thus warping the particle. This doesn't seem to explain wave motion, cherenkov radiation, or why they would ever slow.

Einstein proposed one wave effect to test his general theory of relativity. Einstein predicted that a gravitational redshift of light would occur near high gravity sources based on the equivalence principle. It was very difficult to measure astrophysically. But finally it was measured by Walter Sydney Adams in 1925, and then finally confirmed with accuracy by the Pound–Rebka experiment in 1959. In this test, they measured the relative redshift of two sources situated at the top and bottom of Harvard University's Jefferson tower using an extremely sensitive phenomenon called the Mössbauer effect.

In the Mössbauer effect the narrow resonance absorption for nuclear gamma absorption can be successfully attained by physically immobilizing atomic nuclei in a crystal. The immobilization of nuclei at both ends of a gamma resonance interaction is required so that no gamma energy is lost to the kinetic energy of recoiling nuclei at either the emitting or absorbing end of a gamma transition. Such loss of energy causes gamma ray resonance absorption to fail. However, when emitted gamma rays carry essentially all of the energy of the atomic nuclear de-excitation that produces them, this energy is also sufficient to excite the same energy state in a second immobilized nucleus of the same type.

The result was in excellent agreement with general relativity. This was one of the first precision experiments testing general relativity.

In Einsteinian universe space-time is ruled by gravity which travels in waves. Other theorists posit an actual Gravitron particle. Both seem contradicted and inobservable.

Instead of Gravity as Einsteinian gravity, let's consider an effect of heavy mass on sub-quantum-lattice space-time (or SQL). Einstein did discuss warping of space-time by mass. This is always projected as a bowling ball dropped on a rubber sheet. But it is unsatisfying.

Belief in an underlying SQL can provide perhaps more specifically testable experiments. The effect of the SQL perhaps is something we have come to know as gravity. But the mechanism and explanation in this theory is different.

Let's look at a uniform SQL:

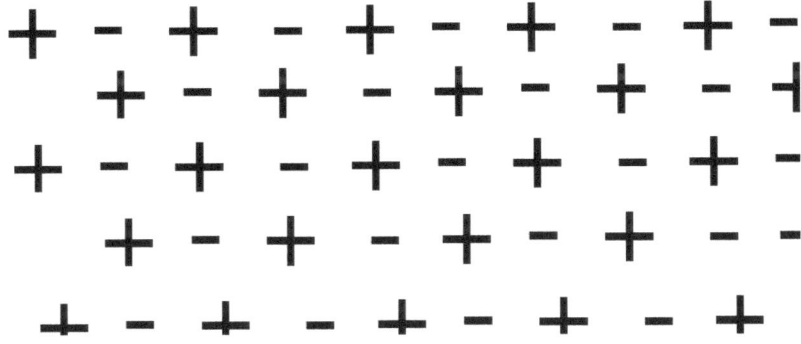

```
+  -  +  -  +  -  +  -  +  -
   +  -  +  -  +  -  +  -  -|
+  -  +  -  +  -  +  -  +  -
   +  -  +  -  +  -  +  -  -
 +  -  +  -  +  -  +  -  +
```

Try to picture it as 3 dimensional not planar. Now lets suppose the sun, a high mass object, resided in the center. Perhaps this warping would be something like:

```
+   -   +    -   +   - + -  +  -
    +  -  +   -   + - +    -  -
+   -   +   -   +   - + -   +  ·
    +  -  +   -   + - +    -   ·
  ,         ,         ,    _  ⊥  -  +
```

Since the SQL is directly involved in rate of wave oscillation and amplitude, then this distortion would increase wave frequency the closer you got to a high mass object.

Similarly, if gravity is effected via a particle, a gravitron, this change in density of the SQL would explain why gravitational effects warp around high mass objects.

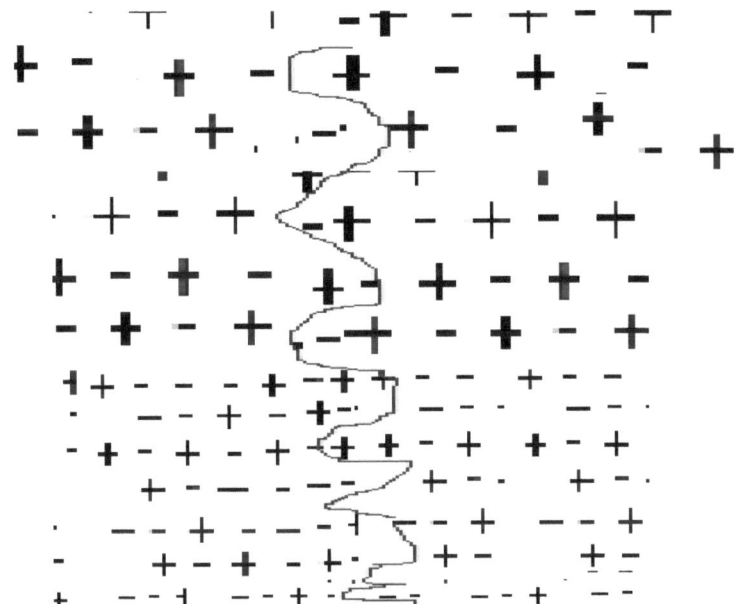

A particle encounters denser and denser SQL space-time as it approaches the black hole until all forward movement is stopped.

As more particles enter, they increase mass, increase the crunching together of the SQL fabric, and become trapped.

Chapter 8 – Some Serious Problems With Gravity and General Relativity

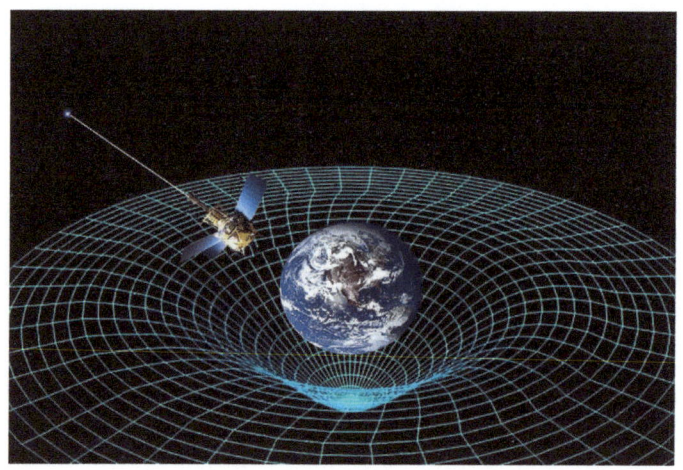

Gravity works because space-time is curved around a mass and then the gravitron particles bounce back and forth. And then some unicorns come out and fly away with pegasus. I don't buy any of it. Hookey. Balderdash. Let's get back to Einstein's theory:

Proceeding on the basis of the experience gained from Maxwell's theory of the electric field, Einstein postulated the existence of a gravitational field that propagates at the speed of light, c, and that will mediate an attraction as closely as possible equal to the attraction obtained from Newton's theory. From the outset it was clear that mathematically a field theory of gravitation would be more involved than that of electricity and magnetism. Whereas the sources of the electric field, the electric charges of particles, have values independent of the state of motion of the instruments by which these charges are measured, the source of the gravitational field, the mass of a particle, varies with the speed of the particle relative to the frame of reference in which it is determined and hence will have different values in different frames of reference. This complicating factor introduces into the task of constructing a relativistic theory of the gravitational field a measure of ambiguity, which Einstein resolved eventually by invoking the principle of equivalence.

Einstein discovered that there is a relationship between mass, gravity and spacetime. Mass distorts spacetime, causing it to curve. Gravity can be described as motion caused in curved spacetime .

The first problem is they call this a "curve" yet no tests for such curvature have ever been performed. NO the effect on light near the sun that really isn't a test of curving only on gravity effect on light. And a particle theory of gravity are they kidding me? It takes eight hours for a photon to reach Pluto and bounce back to the sun. Does this system really work with an eight hour delay?

The lattice model has a very different explanation of gravity. If you picture a 3-dimensional PLUS like a kids Jaxx piece floating there in space, as one component of the lattice, surrounded by it's corresponding positive and negative pieces, what if instead of space-time being curved, it was instead that our universal lattice angled inwards in response to mass. Space time would remain straight, but the lattice would be like a shear effect applying an inward force to items which try to travel along a escape vector. This would happen as a kind of drag effect.

Does gravity then cause the area of space to have a lattice warp shift? What is the mechanism? What if it's a residual effect of object condensation. Perhaps the Sun was once as massive as to extend out to Plutonic space. As it extended inward and cooled it left behind a destructive path in the lattice resulting in angle deformation. Moving through a straight angled lattice an item would move straight. But moving through a angled lattice, curve is induced OR forward motion can be retarded. If you believe in the vibratory model of the universe then the lattice is vibrating itself. In order for movement to be possible it must not bounce into this lattice pointing inward. So you try to jump off the earth but do not get very far as the lattice blocks forward motion then reverses it.

One question however, is how could this follow massive solar scale objects as they path around the galactic center. Wouldn't there be TRAILS of warped lattice with residual gravity? Actually physicists are modeling such things already and we simply need clear experiments proposed.

"We use a weak field (Newtonian) approximation of gravity and consider the gravitational effect from distant,

multiple copies of a large, collapsed (virialised) object today (i.e. a massive galaxy cluster), taking into account the

finite propagation speed of gravity, in a flat, multiply connected universe, and assume that due to a prior epoch of fast expansion (e.g. inflation), the gravitational effect of the distant copies is felt locally, from beyond the naïvely calculated horizon. ... We find that for a universe with a

$T1 \times R2$ spatial section, the residual Newtonian gravitational force (to first order) provides an anisotropic effect that repels test particles from the cluster in the compact direction, in a way

algebraically similar to that of dark energy." - A weak acceleration effect due to residual gravity in a multiply

connected universe, Boudewijn F. Roukema, Stanislaw Bajtlik, Marek Biesiada, Agnieszka Szaniewska, and Helena Jurkiewicz , 5 f´evrier 2008

If we think back to how movement is possible at all, it must be that our initial universe fabric is itself in a state of vibration.

Explaining the Shapiro Effect:

Dr. Shapiro went on to suggest that this new test of relativity theory could be verified by observing the time delay of radar signals returned from the surface of the planets Venus and Mercury. He estimated that the effect of the sun's gravitational field on the radar beam would be to cause a delay of as much as two hundred microseconds (0.0002 seconds) in the round trip travel time of a radar signal returned from a distant planet. The maximum delay would occur at the beam's closest approach to the sun. He went on to explain how, with the knowledge and technology available (in the mid 1960's), such a test could be successfully made to within five to ten percent accuracy using the MIT Haystack radar.

His idea was to bounce radar beams off the surface of the planets Venus and Mercury, and measure the total time it took for the beams to go from the earth to these planets and return. Since the relative positions of the planets and earth are known quite accurately, the expected travel time of the radar beam could be computed with great accuracy as well. His solution of Einstein's equations of relativity indicated that as the radar beam passed closer and closer to the sun, there would be a small time delay. The total time for the radar beam to go from the earth to the planets and back, at the closest approach of the radar beam to the sun, would be increased by 200 microseconds compared to what would be expected if the sun were not there. This is a relatively easy time difference to measure.

the most accurate experiments of the Shapiro effect have been conducted as a result of NASA's Viking project. This program placed unmanned landing craft on the surface of the planet Mars to explore its characteristics. One of the wonderful results of this program, you may recall, was to return colour photographs of the Martian surface. A lesser known part of this program was to leave transponders on the surface of Mars. These transponders respond to radio signals from earth and return, or "*echo*" these signals back to earth, ideal for testing the gravitational time delay. Such controlled signal response from fixed positions eliminates both the random nature of raw radar returns from a planetary surface, and possible orbital variations present when returning signals from the Mariner spacecraft.

As the line of sight between Earth and Mars drew closer and closer to the sun, a measurable excess time delay began to occur. When the line of sight came nearest to the Sun (called *superior conjunction*), the maximum excess time delay occurred — about 200 microseconds as predicted by Shapiro's equations.

Time dilation is another key effect discussed by Einstein. Time would be effected by both velocity and gravity. But the mechanism of this effect is never discussed. How does curved spacetime alter time reference? It's a bit baffling.

But with a lattice model of a vibrationary universe we can see how trying to accelerate through the lattice might induce an effect at the atomic and subatomic level. Whereas at rest the vibration in the atom would be more free to its full expression, at huge acceleration there would be drag/collision effect with the lattice resulting in reduced vibration in the atom with perfect conservation of energy. Stop the acceleration and the atom moves back to normal paths of orbits and particle speeds.

Thus, a picture of why time dilation occurs starts to take form, because time is the fundamental vibration rate of the universe.

Gravitational lensing (light bent by a star seeming to move a distant star into an incorrect position as the path of a high mass object like our sun approaches into the sight line) has been confirmed and once again rejoicing at the general theory of relativity. But mechanism is utterly obscured. Einstein did no better then describing the nature of the universe but not the why or how. The lattice model helps fill in some gaps into this thinking. If high mass objects resulted in the deformation of the lattice (think of 3 dimensional Xs (jacks) now tilted rather than straight) and this is vibrating, as a bundlet of energy like a photonic beam comes across it, it will strive for a path of least resistence and bend.

Gravity suffers many issues similar to the issue of why a hydrogen electron doesn't decay. But on a grand scale consider a binary neutron star. The theory goes gravity waves would increase each others acceleration, yet we observe no energy/matter loss that seems like it would have to be required by such massive forces.

Gravity therefore must be a substrate effect not a produced distance effect. An atom in it's wild flight has a vibration. If you picture a trillion trillion of them there is a concept of unified vibration. Not the vibration like sound or something we humans can feel on our scale, but vibration that is both on a very deep sub quantum scale, and a very large distance scale. This resonance corrupts lattice vibration resulting in what might be seen as structural deformation. Perhaps this deformation is angular or symmetry destroying. If symmetric lattice travel or harmonic travel is smooth, moving through non symmetric, angled, or non harmonic vibration sub quantum lattice changes velocity. It is not clear yet how it can be an attractive force. Maybe like muscle fibers which slide across each other like inch worms with little nubs pushing slightly backwards producing a backwards effect or contraction, trillions upon trillions of quantum lattice vibrations might back push upon themselves, pulling atoms in their wake together. So literally atom by atom or even smaller than that, mass is moved by an effect which is entirely local, without waves or particles, but resultant of mass nearby.

"it has been widely expected for decades now that at a scale of 10^{-33} centimeters, the structure of space-time will cease to be the implacable, smooth 'surface' we use in modern quantum theory and become something quite bizarre. Perhaps an unimaginable froth of mini-worm holes, quantum loops or strings wiggling about in some strange kind of hyperspace with a dozen or more dimensions. At these scales, all particles loose their point line character. All quantum fields reduce to some more complex topological structure as the superstring theorists advocate" - Dr. Sten Odenwald

Chapter 9 – Summary: The lattice model and vibe-atomic model of energy and organization in the universe

If we are to develop a grand unified theory we might come to realize that quantum theory was a miss-step that fit observational data. It happened because of a seduction by the planetary systems that seemed to mirror what happened in the atom then again in sub atomics. It happened because we lack the physics to describe a resonant model of the universe with energy packlets that differentiate based on twist and spin. And I worry very much that we arrive at different "particles" simply because one energy-let spins, vibrates, twists, or tumbles over itself in different ways and at different rates. We have built a descriptive category of actors which is useful, but not theoretical observable or at the foundation of core mechanism.

The end-time model of the universe as lattice, and quantum lattice study in general, are usable theories to get us to begin to think about what might be for our current universe which does NOT exhibit such a structure typically. But it might tell us something about relationship, combination, and decay. Such a lattice model (I don't want to say quantum anymore) distribution of energy in our universe may give rise to these vibration effects which produce the wave of light and the curve of the electron field.

Is this lattice model simply an over-zealous quest for perfection? Like different lattice structures of carbon, we know we achieve ever different properties and hardness. So we are seduced by the notion of putting the blocks in their perfect order.

The lattice vibrationary model helps us understand areas where general relativity simply says an effect occurs yet supplies no mechanism. As such, general relativity was an incomplete theory. The lattice vibrationary model will require new experiments and visions as to discern the difference from general relativity but one issue with this is that curved space time simply is not a testable hypothesis. It is better described as a guess, a mental vision. The vibrationary lattice fills in to explain how such things can occur solving the inherent issue with effect of grand scale distance at speeds of light.

Like many theories before, it asks us to go to a much deeper scale in the universe that all the rest of our atomic and even sub atomic knowledge is simply epi-phenomena. Still, there should be testable results one of them being it allows for effects greater than the speed of light while wave/particle theory does not. (without resorting to some time distortion presumption). As such, it may be a cleaner model. There may be other effects which occur in areas of severe disruption of the lattice, like what may occur around a supernova. As a guess, the effect would be one of chaos and deceleration, and perhaps extreme gravity. This result isn't predicted by general relativity, after a supernova to dwarf reaction GR would predict regular space. So if we find distortions that are large and unexplained in this area, we may indeed have first fingerprints to the lattice model.

This new model focused on alignments, twists of energy, and vibration not in the obscured notion that quantum physics has brought us is needed. Such a model might bring much together in our minds on the true workings of the universe and finally extend from the unfathomably small to unfathomably vast universe. As such, it is much more unified than where we were with Einstein and certainly with the wrong direction of string theory and quantum descriptions being over-extended into areas like gravitrons and other notions of particles as an exchanging force.

Our brains are drawn against a vibrationary model both because of the prior discredited ether notions but also because it simply is much more complex. It brings us back to the true nature of motion in a void and how even movement is possible. The notion that our universe may have a underlying symmetry is somehow a comfort, perhaps something knitted by God, a secret Man was never expected to understand.

My hope is that other researchers who have the mathematics and vision will work on the prediction nature of this model as it disrupts general relativity and begin the long laborious task of evidentiary based science. For me, the hypothesis itself was rewarding.

---- Gianna Giavelli, 2015

www.ingramcontent.com/pod-product-compliance
Lightning Source LLC
Chambersburg PA
CBHW040854180526
45159CB00001B/424